Environmental Aspects, Applications, and Implications of Nanomaterials and Nanotechnology 2018

Topical Conference at the 2018 AIChE Annual Meeting

Pittsburgh, Pennsylvania, USA
28 October - 2 November 2018

ISBN: 978-1-5108-7643-9

Printed from e-media with permission by:

Curran Associates, Inc.
57 Morehouse Lane
Red Hook, NY 12571

Some format issues inherent in the e-media version may also appear in this print version.

Printed by Curran Associates, Inc. (2019)

AIChE
120 Wall Street, FL 23
New York, NY 10005-4020

Phone: (800) 242-4363
Fax: (203) 775-5177

www.aiche.org

Additional copies of this publication are available from:

Curran Associates, Inc.
57 Morehouse Lane
Red Hook, NY 12571 USA
Phone: 845-758-0400
Fax: 845-758-2633
Email: curran@proceedings.com
Web: www.proceedings.com

TABLE OF CONTENTS

Some abstracts in this book refer to images that were not made available by the publisher

527130 Shape Matters: Cr(VI) Removal Using Iron Nanoparticle Impregnated 1-D Vs 2-D Carbon Nanohybrids Prepared By Ultrasonic Spray Pyrolysis

Abstract:

Iron nanoparticles (Fe NPs) are used for treating water contaminated with metals or organic compounds. One-dimensional (1-D) carbon nanotubes (CNTs) and two-dimensional (2-D) graphenes act as useful nanocarbon (NC) supports for Fe NPs by resisting aggregation and enhancing adsorption and redox activity. However, no study showed how shape difference between tubular CNT and planar graphene structures dictates the physicochemical properties and pollutant removal potential of their iron-based nanohybrids. In this work, ultrasonic spray pyrolysis was used to continuously prepare Fe-CNT and Fe-rGO nanohybrids. Both NC shape and Fe/NC ratio influenced Fe NP size, loading, and oxidation states. High Fe content (precursor Fe/NC mass ratio = 2) resulted Fe NPs with diameters of 30.97 ± 7.00 and 24.11 ± 4.42 nm for Fe-CNT and Fe-rGO, respectively; however, low Fe content (Fe/NC = 0.2) provided more uniformly dispersed Fe NPs of 15.65 ± 3.06 and 9.67 ± 1.49 nm, respectively, while unsupported Fe NPs were 285.71 ± 132.42 nm. The USP-derived nanohybrids, for the first time, were used for removal of pollutant, i.e., chromium (Cr(VI)) from aqueous media. Both CNT and rGO provided synergistic effects to significantly enhance Fe NPsâ€™ ability to remove Cr(VI); the effect was more pronounced in Fe-rGO than Fe-CNT and also for low Fe content in both cases. Fe-rGO with low Fe/NC ratio and smallest Fe NPs provided the highest Cr(VI) removal capacity (25 mg/g), which was a multifold improvement over bare Fe NPs and other synthesized nanohybrids (range 7â€“14 mg/g). Overall, 2-D rGO improved contaminant removal capacity of the nanohybrids more than 1-D CNT indicating towards shape effect of NC supports.

529000 Carbon Nanotube-TiO_2 Composites for Photocatalytic Oxidation of Volatile Organic Compounds

Abstract:

Carbon nanotubes (CNTs), when paired with titanium dioxide (TiO2), can increase the charge recombination time of the photocatalyst, improving quantum efficiency. Higher quantum efficiency means excited electrons have a better chance of reacting with adsorbed pollutant molecules which therefore increases the effectiveness of the photocatalyst. CNT/TiO2 photocatalysts have been prepared by adding titanium (IV) chloride to a solution of oxidized CNTs dispersed in H2O2. The photocatalysts have been characterized by X-ray diffraction (XRD), X-ray photoelectron spectroscopy, and scanning electron microscopy (SEM) to determine TiO2 crystal structures, surface morphology and chemical states. The activity of CNT/TiO2 photocatalysts with different amounts of CNTs has been studied by degrading acetaldehyde, a volatile organic solvent representative of organic atmospheric pollutants. The photocatalytic degradation reaction data revealed that the presence of a small amount of CNT additive not only changes the morphology of the composite, improving porosity, but also enhances the activity of the photocatalyst.

531698 Synthesis and Characterization of Analyte-Responsive Nanoparticles for the Detection of Polychlorinated Biphenyls

Abstract:

Polychlorinated biphenyls (PCBs) constitute a class of persistent organic pollutants (POPs) that, as a consequence of their high chemical stability and prevalent production in the early-to-mid 20th century, have accumulated in the environment, posing a significant threat to the health of exposed populations. As part of the efforts to identify and remediate heavily polluted areas, including Superfund sites, easily implemented methods to detect the concentration of PCBs in ground and surface water are necessary to appropriately determine the severity of contamination and designate the sites for quarantine and cleaning. Standard quantification as per Environmental Protection Agency (EPA) guidelines employs gas chromatography (GC) to discriminate and measure the type and concentration of POPs in a given sample, but the technique requires several preparatory extraction procedures that can mar the accuracy of results. GC systems are also time consuming, expensive, and often difficult to operate, altogether making them non-ideal for rapid, field-ready analysis of suspect samples. As a facile alternative, polymeric nanoparticles composed of poly(N-isopropylacrylamide) (PNIPAAm) formed with analyte-responsive domains present a solution-based approach for monitoring the presence of hydrophobic contaminants such as PCBs in aqueous environments. Metrics like turbidity permit visual identification of an analyteâ€™s occupancy, and complementary analysis with ultraviolet-visible spectroscopy (UV-Vis) and dynamic light scattering (DLS) allow for quantification. Successful incorporation of comonomers (e.g., 4-vinylbiphenyl, 4-vinylaniline) into the nanoparticle network is evaluated using Fourier transform infrared spectroscopy (FTIR), and electron microscopy (EM) is used to verify the size distribution of the unswollen nanoparticles. Detection of hydrophobic contaminants with functionalized nanoparticles provides an easily implemented sensing method that is tunable for specific analytes with judicious selection of comonomers and adaptable for development into substrate-functionalized sensors.

532083 Eco-Friendly Fabrication and Characterization of Mechanically Strong, Thermally Stable, Largely Impermeable and Biodegradable Zein-Graphene Oxide Nanocomposites

Abstract:

Eco-friendly fabrication and characterization of mechanically strong, thermally stable, largely impermeable and biodegradable Zein-Graphene Oxide Nanocomposites

Tahrima B. Rouf, Jozef L. Kokini

Department of Food Science, Purdue University, West Lafayette, IN47907

Email: trouf@purdue.edu, jkokini@purdue.edu

Synthetic polymers are polluting the environment at an alarming rate. It has been reported that the "Great pacific ocean garbage patch", which was twice the size of Texas [1], is now thrice the size of France and consists of 79,000 tons of plastic waste [2]. As an eco-friendly alternative, natural polymers have received tremendous attention, due to their nontoxicity, biodegradability, biocompatibility, low cost, non-dependence on petroleum sources, and availability from renewable resources. Zein, the most abundant protein of corn, is a byproduct of ethanol production and is one of the most commonly used plant-based proteins. Due to its unique hydrophobic film forming properties, low cost, abundance and lack of essential amino acids for human consumption, zein is often used in non-food applications [3]. Zein has found application in drug delivery [4–6], coating material for tablets, nanoparticles and nanofibers [7, 8], high strength adhesives [9] and packaging materials [10, 11]. Typically standalone zein proteins often require some chemical or physical modifications, like addition of plasticizers, crosslinker or reinforcing nanofillers to have comparable properties to those of synthetic polymers [10–12]. The objectives of this study were to successfully fabricate zein-graphene oxide (Z-GO) nanocomposite films and characterize the obtained product, for offering an in-depth insight regarding the effect of graphene oxide (GO) nanofiller in the mechanical, thermal, barrier and surface properties of zein. Graphene oxide (GO), is a reinforcing nanomaterial made of single or few layers of carbon atoms, that has reportedly increased mechanical, thermal, barrier, electrical properties of different plant based biopolymers [13]. GO nanosheets were prepared using Tour's method. Z-GO nanocomposites were fabricated using solvent casting technique, by applying GO in concentrations of 0%, 1%, 3%, and 5%, where 70% ethanol was used as a solvent, oleic acid used as a plasticizer and monoglyceride as an emulsifier. The prepared GO nanosheets and the Z-GO nanocomposite films were characterized using TEM, AFM, FT-Raman, FTIR, TGA, as well as mechanical property and water vapor permeability measurements. The TEM of the prepared GO nanosheets showed wrinkled morphology, which resembles crumpled silk veil waves [14] and the transparency of the GO sheets indicated exfoliation to few layer structured graphene material. This effect resulted in obtaining higher surface area and further increased the possibility of achieving good biopolymer/filler interface. The FT-Raman spectra for the multi-layered GO sheets exhibited two characteristic peaks located at 1309 and 1590 cm^{-1}, which can be attributed to D and G bands. The G band is caused by in plane bond stretching vibrations of sp^2 bonded carbon atom and the D band is due to vibrations of the carbon atom of disordered and defected graphite [15]. The prominent D peak in the FT-Raman spectra is an indication of the presence of disorder, which is likely related to both the structural disorder induced by the tears and folding over of the GO sheets as well as the presence of residual oxygen and point defects. Chemistry of bond formation in GO nanosheets, as well as, Z-GO nanocomposite was investigated using FTIR. FTIR spectra of pristine GO shows characteristic peaks at 1373 cm^{-1} (C–O–C vibrations), 1623 cm^{-1} (sp^2-hybridized C=C, in-plane vibrations, as well as vibrations from ketonic species (C=O)), 1070 cm^{-1} (C–OH vibrations), 1737 cm^{-1} (C-OH vibrations from –COOH), and 3207 cm^{-1} (C–OH vibrations from COOH and H_2O) [16]. FTIR of Z-GO nanocomposites showed, the characteristic protein peaks from zein, as well as, formation of a new peak at 1048 cm^{-1} which can be attributed to C-O stretching vibrations from GO. As the nanofiller concentration was gradually increased, the intensity of the new peak also increased, which indicated that GO has been distributed uniformly throughout zein matrix. Surface roughness analysis using AFM showed that Z-GO nanocomposites had reduced roughness compared to control zein films, which is another indication of the uniform dispersion of GO throughout zein matrix and water contact angle measurement showed significant changes in hydrophobic/hydrophilic balance, on addition of GO. TGA results showed that the 50% thermal decomposition temperature increased by 16 ^{o}C, and the 90% thermal decomposition temperature increased by 34°C at only 1.5% GO loading. Our studies also showed significant improvement of mechanical properties as well as a decrease in water vapor permeability for the nanocomposites with only a small loading of GO. The impact of GO on properties of zein films strongly depended on the percentage as well as the preparation technique. This research provides an in-depth analysis of the impact of GO nanofillers on the physicochemical properties of zein biopolymer and aims at providing a comprehensive understanding of the properties of nanocomposites, with an ultimate goal to find novel applications for bio-based food grade materials in the fields of packaging, sensors as well as other biodegradable functional films.

References

1. Sigler M (2014) The Effects of Plastic Pollution on Aquatic Wildlife: Current Situations and Future Solutions. Water Air Soil Pollut 225:2184 . doi: 10.1007/s11270-014-2184-6
2. Lebreton L, Slat B, Ferrari F, et al (2018) Evidence that the Great Pacific Garbage Patch is rapidly accumulating plastic. Scientific Reports 8:4666 . doi: 10.1038/s41598-018-22939-w
3. Turasan H, Kokini JL (2017) Advances in Understanding the Molecular Structures and Functionalities of Biodegradable Zein-Based Materials Using Spectroscopic Techniques: A Review. Biomacromolecules 18:331–354 . doi: 10.1021/acs.biomac.6b01455
4. Hu K, Huang X, Gao Y, et al (2015) Core–shell biopolymer nanoparticle delivery systems: synthesis and characterization of curcumin fortified zein–pectin nanoparticles. Food chemistry 182:275–281
5. Jiang Y-N, Mo H-Y, Yu D-G (2012) Electrospun drug-loaded core–sheath PVP/zein nanofibers for biphasic drug release. International journal of pharmaceutics 438:232–239
6. Zhang Y, Cui L, Che X, et al (2015) Zein-based films and their usage for controlled delivery: origin, classes and current landscape. Journal of Controlled Release 206:206–219
7. Guo HX, Shi YP (2009) A novel zein-based dry coating tablet design for zero-order release. International journal of pharmaceutics 370:81–86
8. He M, Jiang H, Wang R, et al (2017) Fabrication of metronidazole loaded poly (ε-caprolactone)/zein core/shell nanofiber membranes via coaxial electrospinning for guided tissue regeneration. Journal of colloid and interface science 490:270–278
9. Schmidt G, Hamaker BR, Wilker JJ (2018) High Strength Adhesives from Catechol Cross-Linking of Zein Protein and Plant Phenolics. Advanced Sustainable Systems
10. Luecha J, Sozer N, Kokini JL (2010) Synthesis and properties of corn zein/montmorillonite nanocomposite films. J Mater Sci 45:3529–3537 . doi: 10.1007/s10853-010-4395-6
11. Rouf TB, Schmidt G, Kokini JL (2018) Zein–Laponite nanocomposites with improved mechanical, thermal and barrier properties. J Mater Sci 1–16 . doi: 10.1007/s10853-018-2061-6
12. Gezer PG, Brodsky S, Hsiao A, et al (2015) Modification of the hydrophilic/hydrophobic characteristic of zein film surfaces by contact with oxygen plasma treated PDMS and oleic acid content. Colloids and Surfaces B: Biointerfaces 135:433–440 . doi: 10.1016/j.colsurfb.2015.07.006
13. Rouf TB, Kokini JL (2016) Biodegradable biopolymer–graphene nanocomposites. Journal of Materials

Science 51:9915–9945
14. Li M, Wu Z, Ren W, et al (2012) The doping of reduced graphene oxide with nitrogen and its effect on the quenching of the material's photoluminescence. Carbon 50:5286–5291 . doi: 10.1016/j.carbon.2012.07.015
15. Kar P, Sardar S, Liu B, et al (2016) Facile synthesis of reduced graphene oxide–gold nanohybrid for potential use in industrial waste-water treatment. Science and Technology of Advanced Materials 17:375–386 . doi: 10.1080/14686996.2016.1201413
16. Kumar N, Das S, Bernhard C, Varma GD (2013) Effect of graphene oxide doping on superconducting properties of bulk MgB2. Superconductor Science and Technology 26:095008

537429 Gold on Fractal Nanoparticles As Highly Active Surface-Enhanced Raman Scattering Substrates

Abstract:

Surface-enhanced Raman scattering (SERS) is an ultra-sensitive and powerful spectroscopic technique for molecular sensing and detection. The large enhancement of weak Raman signals arises from electromagnetic mechanism (EM) and chemical mechanism (CM). The EM which is the dominant contributor to SERS process, originates from excitation of localized surface plasmons (LSP) at or near gaps, crevices, or sharp tips of plasmonic materials. Whereas, CM is based on charge-transfer between physically or chemically adsorbed analytes and substrate surface. Both enhancement mechanisms require the analyte to be near the optimized plasmonic surfaces. In this work, unique gold nanostructures have been synthesized on para-amino benzoic acid terminated carbon black and fumed silica nanoparticles by using a cationic polyelectrolyte templating technique. This method involves using a cationic polyelectrolyte to provide enough positive charge density to concentrate gold ions concentration at the surface of the template. Fractal particles coated with poly-L-lysine (PLL) were suspended in a tetrachloroauric acid ($HAuCl_4$) solution. PLL enriched the interface with $AuCl_4^-$ anions and ascorbic acid was added as the reducing agent, forming a thin layer of gold (Au) on the fractal nanoparticles. The absorbance spectrum and morphology of the Au nanoparticles was tuned by controlling the $HAuCl_4$ concentration. SERS enhancement of the gold-coated carbon black and fumed silica has also been successfully demonstrated and compared for the detection of 4-nitrobenzenethiol(4-NBT) as well as pyrene as a model polycyclic aromatic hydrocarbon (PAH). The cusps on the fractal particles coupled with the spiky edges of the gold layer significantly enhance SERS signals. Synthetized Au nanostructure were then immobilized on the surface of graphene nanoplatelets (GNPs)/carbon nanotubes(CNTs) composite through vacuum filtration method. Carbon nanomaterials such as GNPs and CNTs and their composites, known as effective absorbents for aromatic molecules, can enhance Raman signals through trapping the organic molecules for the Au surface. Our results suggest that Au nanostructure on fractal nanoparticles compounded with graphene/carbon nanotube composites are new, highly sensitive substrates for sensing and water purification.

516936 Synthesis and Characterization of Cobalt Doped TiO_2–PVA Nanofibers Catalytic Membrane for Photo Degradation of Congo Red from Aqueous Solution

Abstract:

The development of Co/TiO_2–PVA nanofibers catalytic membrane was investigated. The preparation of Co/TiO_2–PVA composite gel was done by mixing 10% PVA polymer solution and 10% Co/TiO_2 which was pre-synthesized by dry impregnation method. However, the nanofiltration membrane which served as a support was fabricated separately using polyethersulfone polymer.

The Co/TiO_2–PVA solution was loaded in a 25 mL syringe. The charged Co/TiO_2–PVA solution was electrospun and formed nanofibers on the surface of the prepared membrane sheet. The optimum electrospinning parameters were found to be 12 cm for distance to the target; 0.33 mm/min for speed of the syringe; and 20 kV for the high voltage.

The Co/TiO_2–PVA nanofibers membrane were characterized by microscopy, spectroscopy and thermal analysis. SEM micrograph showed nanofibers of size ranging from 110 to 500 nm. The presence of cobalt, titanium and oxygen was confirmed by EDS. TGA revealed that the nanomaterial was thermal stable up to 550 ^{o}C in nitrogen. The nanofibers membrane was calcined at 300 ^{o}C for 16 hours in a muffle furnace.

The obtained calcined Co/TiO_2–PVA nanofibrous membrane was tested in the photodegradation of Congo red dye in aqueous solution. The photodegradation was achieved through simulated solar light irradiation and syringe filter prior to UV analysis. It was found that the developed Co/TiO_2–PVA nanofibrous membrane was successful in complete removal of dye at 180 min. The result indicates that our developed Co/TiO_2–PVA nanofibers membrane could be recommended for further heterogeneous catalysis applications

527756 Aerosol Synthesis of Oxygen-Deficient Titania in a Hot-Wall Reactor

Abstract:

Titania is one of the most widely applied oxide materials in daily life. Its high refractive index makes it an excellent white pigment used in paintings, plastics and sunscreens among others. Another attractive property of titania is its photocatalytic activity, enabling it to disintegrate organic substances or even water splitting when exposed to sunlight [1]. During the last years attempts, have been made to increase this catalytic performance by doping with other metals, forming composites e.g. with carbon or inducing oxygen deficiencies [2]. The latter method leads to a partial reduction of the tetravalent titanium cation and the formation of a series of complex titanium oxides called Magnéli phases[3]. Their stoichiometry can be generally written as Ti_nO_{2n-1} (with n between 4 and 9) and compounds like Ti_4O_7 and Ti_5O_9 belong to this group. Within this work, a one-step method to produce well-defined oxygen-deficient titania nanoparticles via a facile gas-phase process is presented and carefully evaluated.

Titanium (IV)-tetraisopropoxide (TTIP) is used as a precursor and supplied via a temperature-controlled bubbler. The steam-laden gas flow is then mixed with hydrogen or air. Furthermore, nitrogen is added to adjust the residence time in the reactor, which consists of a three-zone tube furnace. Consequently, TTIP undergoes decomposition to form titania and afterwards partial reduction. After passing the reaction zone the hot aerosol stream is quenched consequently the particles are deposited onto a fiber filter. The admixture of nitrogen into the system also allows for feeding water vapor into the system by letting it pass through a second temperature-controlled bubbler. This leads to a shift in particle morphology from nanoparticles to submicron spheres.

In this work, the degree of oxygen deficiency, which can be introduced under different process conditions, was investigated. Therefore, different reactor temperatures, residence times and hydrogen contents in the reactor were used to synthesize substoichiometric titania. The XRD results evidence that both higher temperatures and residence times increase the content of Magnéli phases whereas less anatase is produced. The onset of titania reduction occurs above 700 °C. At very high temperatures Ti_4O_7 becomes the predominant phase and, according to Rietveld refinement, or is even present as the single crystalline phase. Regarding the crystallite sizes, values range between 20 and 30 nm depending on the reactor temperature. This is in good agreement with results from scanning electron microscopy which indicates that these particles are single crystals. For elevated reactor temperatures the particles tend to be much larger which is due to strong sintering of the primary particles. Notably, this sintering is much stronger than for unreduced titanium dioxide synthesized at otherwise similar process conditions. The smallest particles are obtained between 700 °C and 800 °C which is in good agreement with [4]. A further reduction of the particle size can be achieved by a lower residence time of the aerosol in the reactor. Under optimized conditions, a decrease of $x_{50,0}$ from 46 nm to 32 nm can be achieved.

All produced powders were stable over several months at ambient conditions and even up to 80 °C.

Depending on their degree of oxygen deficiency, their color varies between white and dark blue. They showed the ability to catalytically oxidize methylene blue upon UV irradiation (380 nm). Here, samples with a high degree of oxygen deficiency generally showed a higher catalytic activity compared to samples produced below the onset temperature of titania reduction. Strongly sintered, pure Ti_4O_7 was relatively weakly active under UV irradiation as it was also found in [5].

Summarizing, a one-step approach for the gas-phase synthesis of titania with a variable degree of oxygen-deficiency was demonstrated. A precise control of the process parameters enables fine tuning of the produced particles' properties and can lead to production of promising materials for future applications in the field of photocatalysis.

This work was supported by the German Research Council (DFG) and the Cluster of Excellence "Engineering of Advanced Materials" (EAM).

[1] A. Fujishima, T. N. Rao, and D. A. Tryk, "Titanium dioxide photocatalysis," *J. Photochem. Photobiol. C Photochem. Rev.*, vol. 1, no. 1, pp. 1–21, Jun. 2000.

[2] X. Chen, L. Liu, P. Y. Yu, and S. S. Mao, "Increasing solar absorption for photocatalysis with black hydrogenated titanium dioxide nanocrystals.," *Science*, vol. 331, no. 6018, pp. 746–50, Feb. 2011.

[3] S. Andersson *et al.*, "Phase Analysis Studies on the Titanium-Oxygen System.," *Acta Chem. Scand.*, vol. 11, pp. 1641–1652, Jan. 1957.

[4] K. Nakaso, K. Okuyama, M. Shimada, and S. E. Pratsinis, "Effect of reaction temperature on CVD-made TiO_2 primary particle diameter," *Chem. Eng. Sci.*, vol. 58, no. 15, pp. 3327–3335, Aug. 2003.

[5] T. Tsumura, Y. Hattori, K. Kaneko, Y. Hirose, M. Inagaki, and M. Toyoda, "Formation of the Ti_4O_7 phase through interaction between coated carbon and TiO_2," *Desalination*, vol. 169, no. 3, pp. 269–275, Oct. 2004.

537570 A Framework for Assessing the Adequacy of Information for Environmental Impact Assessment of Engineered Nanomaterials

Abstract:

Nanotechnology is developing rapidly with engineered nanomaterials (ENMs) now being used in myriad of applications. At the same time there are concerns regarding the potential environmental and health impacts of ENMs and thus the need for environmental impact assessments (EIAs) that are reliant on the available body of information. Conducting EIA for the target ENM requires, at the initial stage of analysis, evaluation of the adequacy of available information regarding: (i) the rate of ENM releases, (ii) environmental exposure concentrations, and (iii) toxicity of the ENM of concern. Tools are then needed to integrate the body of quantitative and qualitative information to arrive at the needed EIA. However, prior to investing significant resources in conducting quantitative EIA, it is critical to first assess the adequacy (i.e., availability and quantitative value) of the existing knowledge base for the specific EIA scenario of interest (e.g., exposure and ENM release scenarios). Accordingly, the present study presents a rigorous decision support tool (DST) that provides: (i) organization of information relevant for EIA into a decision tree (of information flow) suitable for EIA, (ii) representation of the adequacy of information as categories that are quantified in terms of scores signifying the extent of available information, as well as quantification of the uncertainty associated with the unavailability of information, and (iii) aggregation of the information scores that quantify the adequacy of available information for conducting EIA for ENMs. The constructed DST uses evidential reasoning algorithm that handles nonlinearity in the assessments of the range of available information (data extracted and curated from scientific literature) and integrates them into combined scores that reflect whether the available information is adequate (or sufficient) for carrying out the EIA scenario of interest. In order to illustrate the approach, a case study was conducted for three ENMs (i.e., *nano Cu*, *nano ZnO* and *nano* TiO_2) in which baseline conditions were first set based on literature exposure and toxicity data that were curated following a QA/QC screening of 256 publications. Four different scenarios were then designed to explore the impact of deviations due to the attribute weights associated with the different attributes under the different information categories and EIA scenarios. The base case analysis indicated that under baseline conditions, the results suggest that the body of available information for the three selected was adequate for assessing exposure potentials. When considering the various EIA scenarios, it was apparent that there is a significant body of hazard information for Nano ZnO and Nano CuO particularly in terms of in vivo toxicity studies directed at assessing ecological outcomes. The analysis also suggests that the current body of knowledge is sufficient for conducting occupational risk assessment for *Nano* TiO_2, followed by ecological risk assessment for *Nano Cu-CuO*. In all cases, the analysis demonstrated that additional information will be required regarding the fate and transport, environmental release and exposure data. The scenarios considered in the present study demonstrate an approach to Nano-EIA that can serve to support decision analysis in a broader context of ensuring sustainable nanotechnology. In summary, the DST approach enables the analyst to assign specific exposure scenarios and adjust weights, depending on the purpose of the analysis, and visualize the information domain via suitable information maps.

538196 Human Serum Protein Corona Greatly Changes the Interactions between Nanoparticles and a Model Human Erythrocyte (RBC) Membrane

Abstract:

Proteins in physiological fluids rapidly bind to nanoparticles (NPs), forming a protein corona (PC) that ultimately determines the biological identity of a nanoparticle. It is generally believed that the protein corona is responsible for controlling the colloidal stability of NPs, the way in which NPs taken up by cells and how NPs biodistribute and lead to toxicity. Investigating NPs interactions with cell membranes is a critical first step to understanding the extent of biological destabilization when NPs contact cells. This work describes the use of dynamic surface pressure measurements, coupled with fluorescence microscopy to investigate how the NP-cell membrane interactions are modulated by the presence of a protein (human serum albumin), either free in the environment or included in hard and/or soft coronas. Lipid monolayers are formed by depositing mixtures of phosphatidylcholine (PC), phosphoethanolamine (PE) and sphingomyelin (SM) at air-water interface to model a human erythrocyte (RBC) membranes outer leaflet. A suspension of polystyrene (PS) or proteinâ"PS complexes is injected into the subphase, and dynamic changes in surface pressure (Ï€) are monitored to evaluate the extent of NPs binding and the monolayer response. Our results show that PS NPs modify the monolayer phase behavior and cause lipid extraction and hole formation at high monolayer surface pressure (30 mN m^{-1}). Complexation of the PS NPs with proteins promotes their attachment to the lipid monolayers. PS NPs-hard corona (HC) complexes cause lipid extraction at high monolayer surface pressure (30 mN m^{-1}) and lead to the modification of the monolayer phase behavior with hindering of growth of solid domains. NPs-soft corona (SC) complexes, thus far not considered important for such interactions, induced greater changes in monolayer structure than the NPs alone at the same concentration. Compared to NP-HC complexes, NP-SC complexes yield more disordered structure in monolayers with less cohesion between the lipid molecules.

527241 Membrane Lipid Asymmetry Regulates Nanoparticle-Induced Cell Membrane Damage in Red Blood Cells

Abstract:

Over the past decade, there has been increasing interest in understanding the mechanisms through which engineered nanomaterials (ENM) disrupt the plasma membrane of mammalian cells. However, despite a number of mechanistic studies, primarily using vesicles as cell membrane models, the mechanisms of ENM-induced plasma membrane damage have remained elusive. This lack of understanding is in part due to the structural complexities of the cell plasma membrane. In particular, membrane asymmetry, the fact that the cell plasma membrane has different lipids on the exofacial leaflet (i.e. looking away from the cytoplasm) compared to the cytofacial leaflet (i.e. looking toward the cytoplasm) has been completely overlooked in previous studies of ENM-membrane interactions. The present study focused on the role of individual plasma membrane leaflets in regulating nanoparticle-plasma membrane interactions. Specifically, the effects of plain, carboxyl-, amine-, and polyethylene glycol(PEG)-modified fluorescent silica nanoparticles (50 nm) on the integrity of vesicles mimicking the exofacial (V_{exo}) and cytofacial leaflets (V_{cyto}) of the plasma membrane of red blood cells was examined and compared with ENM-induced cell death in red blood cells (hemolysis).

Vesicle integrity was studied by encapsulating the self-quenching fluorescent probe, carboxyfluorescein, in vesicles and studying its leakage after exposure to 0.0001-0.01 g/L of ENMs at 37 °C. Confocal fluorescence microscopy performed on giant unilamellar vesicles (GUVs) was used to monitor nanoparticle effects on vesicle morphology. Forster Resonance Electron Transfer (FRET) experiments were employed to examine particle localization at the membrane. Finally, ENM-induced hemolysis was evaluated by measuring the absorbance of hemoglobin released from red blood cells after incubation with 0.01 g/L of nanoparticles at 37 °C.

Nanoparticle interactions with V_{exo} and V_{cyto} vesicles were drastically different. V_{exo} vesicles showed significant leakage after exposure to plain and amine-modified particles, but were not disrupted by carboxyl-modified and PEGylated particles. Conversely, none of the particles caused significant leakage in V_{cyto} vesicles as evidenced by a lack of significant leakage. In agreement, FRET experiments revealed significant localization of plain and amine-modified particles, but not carboxyl-modified and PEGylated particles, at the surface of the V_{exo} vesicles. Only minor localization with the V_{cyto} vesicles was observed for all particles. Similarly, GUV images indicated visual disruption of V_{exo} vesicles by plain and amine-modified ENMs. Importantly, hemolysis of red blood cells by ENMs was consistent with leakage assays using V_{exo} vesicles. Hemolysis was observed after the incubation of cells with plain and amine-modified particles, but not after incubation with carboxyl-modified and PEGylated particles. In conclusion, these results suggest that artificial vesicles mimicking the exofacial leaflet of the cell plasma membrane predict ENM-induced membrane damage in red blood cells and could have important implications in predicting the cytotoxicity of ENMs.

537793 Assessing Organomodified Nanoclay Pulmonary Toxicity across Its Life Cycle Using Integrated *in Vitro / In Vivo* Approaches

Abstract:

Incorporation of organomodified nanoclays (ONC), montmorillonite coated with different quaternary ammonium compounds, into different types of organic polymers continues to fuel expansion of nanoclay composite technologies. Based on projected widespread use in numerous industrial sectors and commercial products, recent forecasts suggest increased airborne occupational exposures are likely to occur; however, little is known about ONC pulmonary health risks along its life cycle. Here, we characterized physicochemical properties of several different ONCs across their life cycle including as-produced, breakdown products during nanoclay composite machining, and incinerated by-products. We used an integrated *in vitro / in vivo* approach to assess pulmonary toxicity following ONC exposure to compare human macrophage, epithelial, and fibroblast cell model responses to animal model responses to identify key cellular events to help understand mode of action and potential adverse outcomes. By using high-throughput *in vitro* techniques, including multiplex fluorescent high content screening and electrical cell impedance sensing, pristine uncoated nanoclay (UC) exposure elicited increased membrane damage, mitochondrial dysfunction, and apoptosis in epithelial cells while pre-incinerated ONCs (CC) primarily exhibited an apoptotic and necrotic cell death resulting in a loss in epithelial monolayer integrity. Next, UC caused inflammasome activation, cell membrane damage, and apoptosis initiation in macrophages while CC caused large-scale macrophage necrosis with no indication of inflammasome activation. Incinerated ONC byproduct exposure caused acute toxicity, loss of epithelial monolayer integrity, and inflammation activation in macrophages only at high doses. Both UC and CC exposure to fibroblasts indicated their pro-fibrotic potential with increased proliferation, collagen, and reticular fiber deposition. *In vivo* exposure in an animal model indicated that pre- and post-incinerated ONCs caused a low grade, persistent inflammation with pro-fibrotic signaling at Day 28 post-exposure, which correlated with key events identified in the *in vitro* models. In summary, presence of organic coating and incineration status influences potential mode of action of the pulmonary response. The ONC coating protected against an acute silica-associated inflammatory response, but caused membrane damage, low grade inflammation, and a stimulation of extracellular matrix proteins associated with fibrosis. Current research efforts include composite synthesis, generation, and characterization of nanoclay-enabled composite dusts following controlled machining experiments to understand physicochemical characteristics of released particles and how their exposure affects our high-throughput *in vitro* screening models. Linking physicochemical properties to adverse effects in both *in vitro* and *in vivo* models allows for rapid assessment and information to inform and improve prevention-by-design strategies.

527107 Interactions and Toxicity of Next Generation Graphene-Metal Nanohybrids at the Pulmonary Interfaces: Influence of Emerging Physicochemical Properties

Abstract:

Carbon-metallic nanohybrids are emerging next generation nanomaterials with improved properties and multifunctionalities, and are being pursued for different applications including biomedical, electronic, energy, and environmental fields. However, their potential inhalation exposure and toxicity to humans have not been investigated. This becomes important since these emerging nanohybrids have altered and unique physicochemical properties compared to their parent materials. Thus, the question remains: if the conjugation of the two or more component materials will result in antagonistic or synergistic toxicological consequences. To answer this question, we performed cytotoxicity studies on nanohybrids made from depositing iron nanoparticles onto reduced graphene oxide nanosheets (rGO-Fe), and compared with their parent materials. rGO-Fe nanohybrids were synthesized in house and characterized for physicochemical properties using high resolution transmission electron microscopy, x-ray diffraction, Raman spectroscopy, and thermogravimetric analysis. Moreover, their colloidal stability and sedimentation characteristics in the cell-culture media were determined. Human bronchial epithelial cell line (BEAS-2B) was exposed to the nanohybrids and component materials. Interesting enough, the cell viability for rGO-Fe was found to be lower than for Fe nanoparticles, but higher than for GO. While GO in the cells were distributed evenly, Fe and rGO-Fe nanohybrids accumulated near the nucleus. GO was uptaken slower than Fe and rGO-Fe due to GOâ€™s higher colloidal stability. The trends for cell membrane damage (i.e., LDH leakage), extracellular ROS, and intracellular ROS generation were similar to the cell toxicity results i.e., GO>rGO-Fe>Fe. In all cases, a positive dose-response relationship was established. Scanning electron microscopy suggested that the cellular interaction was dominated by the iron nanoparticles where the cellular toxicity was dominated by the GO counterpart. This suggests the complex interplay of the properties of metal and graphitic nanostructures in their hybridized form causes significant alterations of the toxicity towards pulmonary interfaces.

532416 Unraveling the Role of Nitrogen in the Biological Activity of Nitrogen-Doped Graphene

Abstract:

For decades, the nanotechnology community has grappled with the balance of advancing applications and considerations of potential human health and environmental implications of engineered nanomaterials (ENMs). Research on the biological implications of metal and non-metal ENMs has put into perspective with the relative magnitude of inherent hazard of different ENMs, environmental considerations that influence ENM behavior and toxicity, and the ability to manipulate properties to influence interactions at the bio-interface. Despite significant advancements in our understanding of ENM implications, we still lack the ability to translate governing principles between and within classes of ENMs, particularly carbon-based nanomaterials. The recent and rapid emergence of graphene has motivated our research to uncover the governing mechanisms of graphene interactions at the bio-interface and further, establish design guidelines to tailor graphene properties. Building on our previous research findings for graphene oxide, the current project focuses on nitrogen doped graphene (N-graphene). While nitrogen doping is an important approach to modulate the electronic and chemical properties of graphene intended to broaden its applications, very little is known about the influence of this heteroatom on the biological activity of graphene. This work investigates the biological activity of a N-graphene material suite systematically prepared using facile and ecofriendly synthesis methods. The degree of nitrogen doping and chemical state of nitrogen (i.e., pyridinic, pyrrolic, graphitic, and oxidized N) in doped graphene is tailored by thermal annealing under different temperatures and quantified using X-ray photoelectron spectroscopy. The bioactivity of the prepared materials is quantified as the inactivation of a bacterial model organism and the propensity to oxidize the intracellular antioxidant, glutathione. Given the importance of oxidative stress and electron transfer in mechanisms of toxicity, the materials electronic properties are evaluated using electrochemical techniques that enable the investigation of oxygen adsorption, reactive oxygen species production, and heterogeneous electron transfer. Several complementary characterization techniques are employed to evaluate the consequential properties after nitrogen doping such as sheet dimensions, aggregation state, surface area, and structural defects.

529808 Oxidative Potential and Cellular Oxidant Production from Biomass Burning Aerosol

Abstract:

Exposure to particulate matter (PM) is a leading global health risk and has been associated with increased cardiopulmonary morbidity and mortality. Toxicological studies suggest that PM-induced reactive oxygen and nitrogen species (ROS/RNS) production may be a possible mechanism by which PM exposure leads to adverse health endpoints. Recent studies on the relative toxicities of different aerosol subtypes reported high relative toxicity for biomass burning aerosol in terms of oxidative potential.

Here, we present chemical oxidative potential and cellular oxidant production measurements from biomass burning aerosol collected in the southwestern part of the Brazilian Amazon region, where biomass burning emissions have drastically changed the composition of PM and have been shown to negatively impact the surrounding population. Samples were collected during the dry and wet seasons (n = 41). Dithiothreitol (DTT) was used to assess the oxidative potential or concentration of redox-active species present in the sample. Murine alveolar macrophages were exposed to PM extracts for 24 hrs, after which ROS/RNS production was measured. For all PM samples, a significant positive correlation was observed between ROS/RNS and concentrations of monosaccharide anhydride, such as levoglucosan. Exposure to pure levoglucosan was found to be non-toxic. Despite a lack of causation, levoglucosan may serve as an indicator of biomass burning health effects. ROS/RNS estimations were also calculated based on polycyclic aromatic hydrocarbon (PAH) concentrations and their respective ROS/RNS activity. The discrepancy between these estimations and ROS/RNS measured for the water-soluble hydrophobic fraction of PM suggests that concentration addition may not be an applicable mixture model for PM health effects.

537464 Brain Imaging Probes Elicit Microglial Inflammatory Responses and Induce Cellular Morphological Changes

Abstract:

Microglia are the resident immune cells of the brain that number in the billions, and are central in maintaining the homeostasis of the neuronal environment through states of activation and quiescence.[1] Microglia mediate inflammatory responses to foreign and pathogenic material, which can elicit downstream neurotoxicity and neurodegeneration. However, despite the broad-scale implementation of fluorescent probes to query neuron function, little attention has been paid to probe interactions with, and effects on the brain microglia. Probes such as calcium dyes,[2] synthetic nanosensors,[3] and voltage probes,[4] are utilized ubiquitously in neuroscience to study brain structure and neurochemical activity. Herein, we demonstrate the effects of several brain chemistry probes on a SIM-A9 mouse microglial cell line, with particular emphasis on synthetic nanoparticles utilized to image brain chemistry. We show that single-walled carbon nanotube (SWNT)-based dopamine nanosensors are uptaken by, but are not cytotoxic to, SIM-A9 mouse microglia. The activity of SIM-A9 mouse microglia is next quantified with quantitative PCR and with an immune proteomics screen to monitor mRNA and protein expression and secretion levels of common inflammatory cytokine biomarkers. We show that SIM-A9 co-incubation with dopamine nanosensors induces a slight increase in expression of inflammatory cytokines TNF-α, iNOS, IL-1β, and IL-6. The observed nanosensor-induced cytokine upregulation is significantly lower than that caused by lipopolysaccharide stimulation. Live-cell imaging results further reveal that SIM-A9 interactions with SWNT nanosensors elicit a significant change in SIM-A9 cell morphology towards an elongated and ramified state, an *in vitro* indicator of a quiescent rather than inflammatory microglial state that halts cell proliferation. These contrasting effects indicate that nanosensors affect multiple microglial signaling pathways, possibly through a non-canonical immune response. We also discuss how common passivation techniques, namely PEGylation, can mitigate microglial inflammatory responses to brain imaging probes. PEGylated nanosensors show lower uptake into SIM-A9 cells, a reduced inflammatory response, and partial mitigation of SIM-A9 morphological change. Our results motivate greater attention to the effects of ubiquitously utilized neuro-imaging probes on the brain microglia.

References

1. Kierdorf, K. & Prinz, M. Factors regulating microglia activation. *Front. Cell. Neurosci.* **7,** 1–8 (2013).
2. Grienberger, C. & Konnerth, A. Imaging Calcium in Neurons. *Neuron* **73,** 862–885 (2012).
3. Beyene, A. G., McFarlane, I. R., Pinals, R. L. & Landry, M. P. Stochastic Simulation of Dopamine Neuromodulation for Implementation of Fluorescent Neurochemical Probes in the Striatal Extracellular Space. *ACS Chem. Neurosci.* **8,** 2275–2289 (2017).
4. Peterka, D. S., Takahashi, H. & Yuste, R. Imaging Voltage in Neurons. *Neuron* **69,** 9–21 (2011).

537859 Design and Characterization of a New, Portable *in Vitro* Exposure Cassette with Real-Time Monitoring for Aerosol Measurements

Abstract:

The toxicology of aerosols in occupational settings is often performed through collection of particles on a filter prior to reconstitution into cell culture media which can alter the biological effects. Current *in vitro* exposure systems require additional instruments to control temperature and humidity making the system bulky and difficult to take to the field. The Portable *In Vitro* Exposure Cassette (PIVEC) has been designed for personal monitoring, characterized using varying sized copper nanoparticles, tested with alveolar cells, and set-up to incorporate a real time monitor. Three differently sized copper nanoparticles were dispersed into a dry aerosol and measured gravimetrically and on a number concentration basis to determine the deposition efficiency of the PIVEC. A549 cells, a human alveolar adenocarcinoma epithelial line, were grown at the air-liquid interface and exposed to dry copper nanoparticle aerosols. Oxidative stress and cell viability were monitored post-exposure within the first 24 hours via the production of reactive oxygen species and the release of lactate dehydrogenase. The deposition efficiency ranged from 0.5% to 18% depending on method of analysis and size of particle. The PIVEC has been characterized for nanoparticles from 40-800 nm in size with preliminary toxicology testing. Oxidative stress increased over 4-fold from initial measurements within the first two hours post exposure, however there was no significant difference in cell viability at the four hour time point at deposited doses up to 1.63 mg/cm^2. Inclusion of an enzyme-based biosensor to perform real-time monitoring is currently ongoing. The biosensor is functionalized with cytochrome c to measure reactive oxygen species, including hydrogen peroxide and super oxides, through electrochemical detection during aerosol exposures. The PIVEC is a unique device, designed to monitor aerosols using air-liquid interface *in vitro* techniques including a real-time monitor for oxidative stress.

517905 Surface Chemistry Toxicity Parameters Associated with Combustion Produced $PM_{2.5}$ by in Vitro Assays

Abstract:

Introduction: The International Agency for Research on Cancer (IARC) has labeled diesel exhaust as carcinogenic within class 1. While the mechanism(s) by which soot causes the adverse health effects are not known, a great deal of these harmful effects relate to its ability to cause oxidative stress. Thus, oxidative potential, expressed through reactive oxidative species concentration, can be used as a good estimate for its reactivity and toxicity. Accordingly, physical structure and surface chemistry become surrogate measures of its oxidative potential as together they determine the redox properties and polar/acidic character. Based on this premise, we are testing the role and impact of soot structure and surface chemistry upon interaction with bronchial and alveolar epithelial cells. The proposed projectâ€™s goal is the identification of toxicity, oxidative, and pro-inflammatory factors in combustion produced soot arising from using alternative fuels, by studying the particles directly *instead of studying compounds adsorbed on (and removed from) the soot particles*. Model soots with tailored surface chemistry, and specific particle physical structures are being tested for toxicity, oxidative stress markers, effects of lung inflammation and and signaling pathways using cell culture bioassays, as well as protein carbonylation and DNA damage marker assessment.

Methods: An epithelial cell line (BEAS-2B) was incubated with lab-generated carbon (nascent, nitric acid-treated, and ozone-treated carbon) for 6 hr. and 24 hr. incubation times, at different concentrations (0-100 Î¼g/mL). After incubation, we measured cell viability using the MultiTox-Fluor Multiplex Assay. In addition, we extracted RNA and protein from cells to evaluate expression of inflammatory cytokines (IL-1b, IL-6, IL-8) and genes related to the inflammatory response and oxidative stress (TLR4, CCL2, MMP1, and NRF2).

Results: Treatment of carbon particles with nitric acid and ozone resulted in differential oxygen and carboxylic acid content in the $PM_{2.5}$ surface. In BEAS-2B cells, an inverse relationship of cell viability and $PM_{2.5}$ treatment concentration was observed at both time points assessed. Of the three $PM_{2.5}$ preparations, BEAS-2B cells exposed to nitric acid-treated carbon resulted in the largest decrease (89% and 96%) of cell viability after 6 hr. and 24 hr., respectively, indicating that both carboxylic acid and oxygen content of the particle surface strongly contribute to $PM_{2.5}$ toxicity. Additional results showed an increase in gene expression of the inflammatory cytokines IL-1b, and IL-6, and the oxidative stress markers TLR4 and NRF2 at higher concentrations of $PM_{2.5}$ exposure (25-100 Î¼g/mL).

Conclusion: In summary, exposure of different types of particulate matter to lung epithelial cells increases inflammatory responses, with different effects associated with soot components such as particle surface chemistry. Future experiments will determine how functional groups in the $PM_{2.5}$ surface chemistry affect the activation of the observed inflammatory response, and decrease in cell viability of lung epithelial cells.

540262 Effect of Combustion Particle Size on Pathologically Important Responses in Lung Cells

Abstract:

Although combustion-derived particulate matter(cdPM) is a significant contributor to fine and ultrafine PM levels, our understanding of its health effects is complicated by cdPMâ€™s complex nature: a dynamic mixture of particles and condensed material with different sizes, shapes, and chemical compositions that is generated from a range of sources. Furthermore, several confounding factors make comparing results from various studies of PMâ€™s adverse effects difficult including, for cdPM, fuel composition, the age of the combustion device, combustion conditions, lubricants, and sampling methods. For example, experimental and epidemiological studies about the effect of various PM size fractions (PM_{10}, $PM_{2.5}$, and PM_1) provide somewhat conflicting results. As particle size diameter decreases, its relative surface area increases, which in turn increases the bioavailability of atoms or molecules and may lead to increased pro-inflammatory effects. This study focuses on directly testing the effect of cdPM size on biological responses in lung cells.

cdPM with consistent properties is generated from a premixed flat-flame burner combusting a jet-fuel surrogate at an equivalence ratio of 2.5. A nitrogen shroud and a glass housing minimize atmospheric effects. cdPM consistency is verified by examining the particle size distribution and cdPM composition. The sample is then diluted, thermally denuded and directly collected in DI water by using an impinger. The cdPM sample is size segregated by a flow field flow fractionation (FlFFF) into three size ranges (100-150 nm, 150-200 nm and 200-300 nm). The different size fractions are studied for their effect on cell viability, cellular uptake, inflammatory response, cytochrome P450 (CYP) 1A1 mRNA induction. The biological outcomes are evaluated in A 549 human alveolar basal epithelial cells and in THP-1 cells, a pro-monocytic cell line that is differentiated to pulmonary macrophage-like cells. Although all size fractions of cdPM induced a significant inflammatory response compared to the control in both cell lines, the preliminary results indicated that the smallest particle fraction caused 50% more inflammatory response than the largest particle fraction. This suggests that particle size/surface area is an important factor in understanding the links between combustion particle physicochemical properties and health effects.

525800 Toxicological Screening of Metal Oxide Nanoparticles in Liver Context Demonstrates Apoptosis in Hepatocytes *Versus* Pyroptosis in Kupffer Cells

Abstract:

With growing applications of nanotechnology in different fields of science and technology, engineered nanomaterials (ENMs) could enter into human body through different exposure scenarios. The liver is one of the major targets of ENMs as part of the mononuclear phagocyte system (MPS), either as a result of particle uptake and spread from primary exposure sites (*e.g.*, lung and gastrointestinal tract) or systemic administration of therapeutic and imaging nanoparticles. Therefore, it is important to thoroughly investigate how ENMs interact with different cellular components of the liver, and recognize the potential adverse impacts that these interactions could introduce. In this study, we selected 29 metal oxide nanoparticles, some of which are commonly used in consumer products, and performed a comparative analysis of the toxicological impact of these nanoparticles (NPs) in transformed or primary Kupffer cells (KCs) and hepatocytes. We not only observed differences between KCs and hepatocytes, but also differences in the toxicological profiles of transition-metal oxides (TMOs, *e.g.*, Co_3O_4 and Mn_2O_3) *versus* rare-earth oxide (REO) NPs (*e.g.*, Gd_2O_3 and La_2O_3). While pro-oxidative TMOs induced apoptotic cell death in both cell types via activation of caspases 3 and 7, REOs induced lysosomal damage and NLRP3 inflammasome activation, followed by caspase 1 activation and pyroptotic cell death in KCs. Pyroptosis was accompanied by cell swelling, membrane blebbing, significant IL-1Î² release, and increased membrane permeability, which could be reversed by knockdown of the pore forming protein involved in pyroptosis, gasdermin D. These features were not seen in hepatocytes, and REOs induced caspases 3 and 7 activation in this cell type leading to apoptosis. Further investigation of the cytotoxic effects of REO NPs in other phagocytic cell lines such as J774A.1 and RAW 264.7 cells as well as bone marrow-derived macrophages (BMDMs) also demonstrated features of pyroptosis and increased IL-1Î² production. These findings not only demonstrate important mechanistic considerations that can be used for safety evaluation of metal oxides, including commercial products that are developed from these materials, but also provide useful information for safer material design.

AUTHOR INDEX

AIChE
120 Wall Street, FL 23
New York, NY 10005-4020

ISBN 978-1-5108-7643-9

CURRAN ASSOCIATES INC.
proceedings.com